332.4
Mol   Molter, Cary

A quarter = 25 ¢

26242

SandCastle

Dollars & Cents

A Quarter = 25¢

Carey Molter

Consulting Editor, Monica Marx, M.A./Reading Specialist

Publishing Company

Published by SandCastle™, an imprint of ABDO Publishing Company, 4940 Viking Drive, Edina, Minnesota 55435.

Credits
Edited by: Pam Price
Curriculum Coordinator: Nancy Tuminelly
Cover and Interior Design and Production: Mighty Media
Photo Credits: AbleStock, Comstock, PhotoDisc, Rubberball

Library of Congress Cataloging-in-Publication Data

Molter, Carey, 1973-
     A quarter = 25¢ / Carey Molter.
        p. cm. -- (Dollars & cents)
     Includes index.
     Summary: Explains what a quarter is, how it compares to other coins, and how many quarters are needed to purchase different items.
     ISBN 1-57765-899-2
     1. Money--Juvenile literature. [1. Money.] I. Title: Quarter equals twenty-five cents. II. Title. III. Series.

HG221.5 .M658 2002
332.4'973--dc21

                                                                                    2002071192

SandCastle™ books are created by a professional team of educators, reading specialists, and content developers around five essential components that include phonemic awareness, phonics, vocabulary, text comprehension, and fluency. All books are written, reviewed, and leveled for guided reading, early intervention reading, and Accelerated Reader® programs and designed for use in shared, guided, and independent reading and writing activities to support a balanced approach to literacy instruction.

Let Us Know

After reading the book, SandCastle would like you to tell us your stories about reading. What is your favorite page? Was there something hard that you needed help with? Share the ups and downs of learning to read. We want to hear from you! To get posted on the ABDO Publishing Company Web site, send us email at:

sandcastle@abdopub.com

SandCastle Level: Beginning

This is a quarter.

A quarter is a coin.

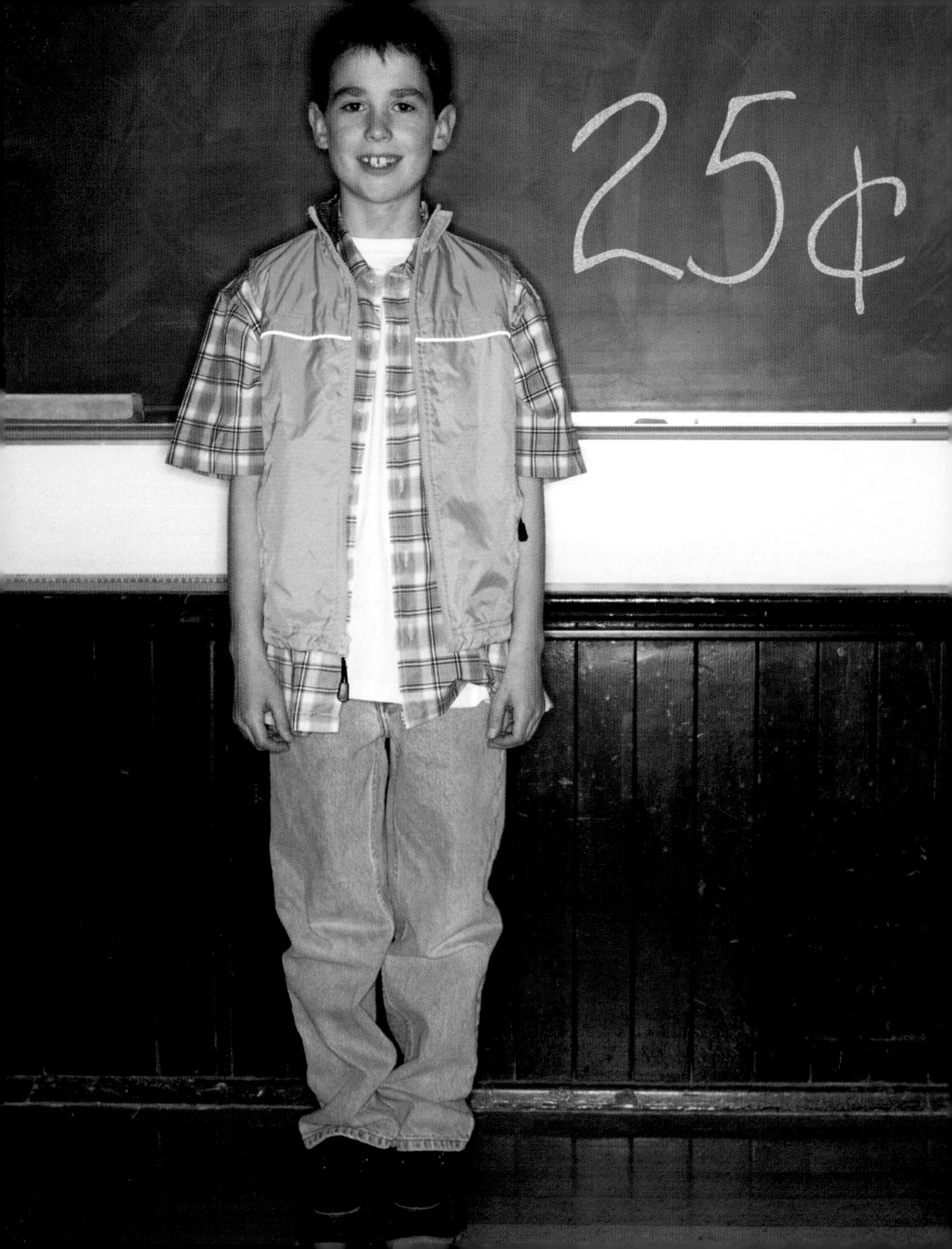
25¢

One quarter is the same as twenty-five cents.

This is how to write twenty-five cents.

25¢

One quarter is the same
as twenty-five pennies.

One quarter is the same
as five nickels.

One dollar is the same
as four quarters.

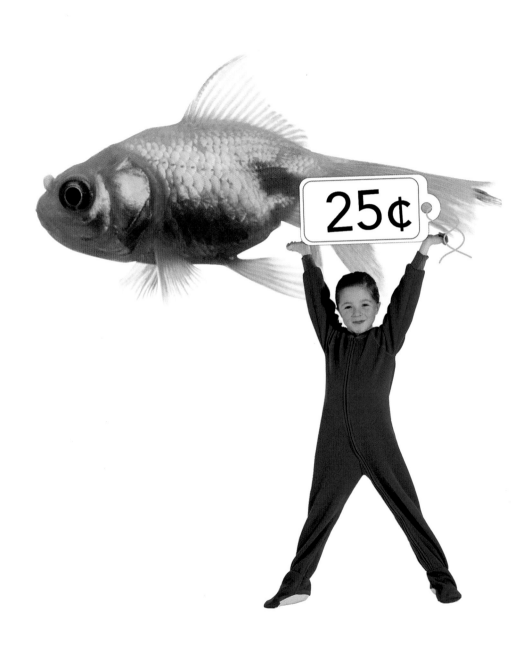

This fish costs 25¢.

That is one quarter.

This gumball costs 50¢.

That is two quarters.

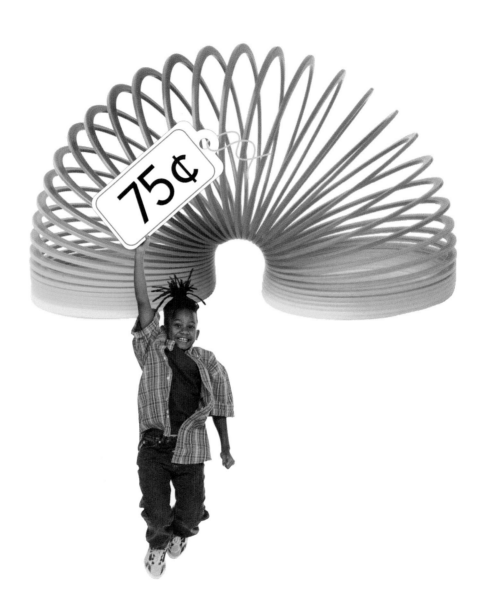

This toy costs 75¢.

That is three quarters.

This ice-cream cone
costs one dollar.

How many quarters
is that?

(four)

Picture Index

fish, p. 15

one dollar, p. 13

quarter, pp. 3, 5, 7,
9, 11, 15

twenty-five cents, p. 7

More about the Quarter

President George Washington

Year the coin was made

Where the coin was made. The letter *D* means it came from Denver, Colorado

E pluribus unum means one out of many

The eagle stands for loyalty to America

Ridged edges

23

About SandCastle™

A professional team of educators, reading specialists, and content developers created the SandCastle™ series to support young readers as they develop reading skills and strategies and increase their general knowledge. The SandCastle™ series has four levels that correspond to early literacy development in young children. The levels are provided to help teachers and parents select the appropriate books for young readers.

Emerging Readers
(no flags)

Beginning Readers
(1 flag)

Transitional Readers
(2 flags)

Fluent Readers
(3 flags)

These levels are meant only as a guide. All levels are subject to change.

To see a complete list of SandCastle™ books and other nonfiction titles from ABDO Publishing Company, visit **www.abdopub.com** or contact us at:

4940 Viking Drive, Edina, Minnesota 55435 • 1-800-800-1312 • fax: 1-952-831-1632